NASA

BEHIND-THE-SCENES SNIPPETS

JAMES A. HENRIE

NASA

BEHIND-THE-SCENES SNIPPETS

An anecdotal story

TATE PUBLISHING
AND ENTERPRISES, LLC

Published by Tate Publishing & Enterprises, LLC
127 E. Trade Center Terrace | Mustang, Oklahoma 73064 USA
1.888.361.9473 | www.tatepublishing.com

Tate Publishing is committed to excellence in the publishing industry. The company reflects the philosophy established by the founders, based on Psalm 68:11,
"The Lord gave the word and great was the company of those who published it."

Cover design by Joseph Emnace
Interior design by Manolito Bastasa

Published in the United States of America

ISBN: 978-1-68118-617-7
Biography & Autobiography / Personal Memoirs
15.01.27

CONTENTS

Dr. Wernher von Braun & Jim Henri

NASA-Marshall Space Flight Center

Huntsville, Ala.

INTRODUCTION

"Mr. President, would you please sit down?"

The following are recollections of some of the more exciting, unusual, and sometimes bizarre events of my professional career with the National Aeronautics and Space Administration (NASA) during Project Apollo—the US manned lunar program in the 1960s.

My service with NASA began in 1962, when I had the good fortune of being employed as a public information specialist at the NASA–Michoud Assembly Facility (MAF) in eastern New Orleans. Prior to this, I was a public relations specialist with the Sheraton hotel corporation in New Orleans. This association with Sheraton provided me with an opportunity to meet the NASA team that came to New Orleans to usher in the production of the rocket launch vehicles to be used in the Apollo program. The Michoud facility was chosen as the site for construction—by the Chrysler Space Division and the Boeing Company—of the mammoth first stages of the *Saturn* launch vehicles. The *Saturn* launch vehicles were to be the transporters of manned spacecraft to the moon within the 1960 decade.

That meeting of the initial NASA team resulted in my being employed by NASA at Michoud. My service with NASA during this time provided opportunities for me that at the time seemed unimaginable. Recall, this was the time when our nation was reeling from the throes of Russian-made satellites orbiting Earth for the first time ever.

The following are snippets of some of the more interesting events, some pleasant and some not.

THE LOWLY MOSQUITO HAMPERS US SPACE PROGRAM

During the early 1960s as the MAF was beginning to establish itself as one of the nation's leading space launch vehicle manufacturing sites, a massive, nation-wide employee recruiting program was begun. The Chrysler Space Division and the Boeing Company began their efforts to recruit engineers, scientists, and administrative-type employees throughout the country.

As this time was the beginning of the Apollo program, national media attention was focusing on what was then called the "Space Crescent," which included NASA installations in Houston, TX; New Orleans, LA; Hancock County, MS; Huntsville, AL; and Cape Canaveral, FL.

One of the more prominent science journalists at the time was Walter S. Sullivan, Jr., science writer for the *New York Times*. Mr. Sullivan planned to include the MAF in his articles as a major member of the NASA Space Crescent.

Mother Nature who seemed to enjoy playing environmental games from time to time presented the most unusual opportunity that apparently was too good to let pass. Adverse environmental conditions in south Louisiana had developed to a point that resulted in one of the largest mosquito invasions ever to hit the New Orleans area. The invasion was so severe that newspapers, television, and radio continually ran stories about how area citizens could best protect themselves from the blood-thirsty critters, stories that included suggestions about type and color of clothing that should be worn, cautions about the use of perfumes and colognes, and what additional measures people could take to better protect themselves.

I can remember, for example, at the end of the workday, we would run as fast as we could from our office building to our automobiles, hoping to avoid the imminent mosquito stings we had to face.

Although Walter Sullivan's objective was to write about the MAF and its role in the space program, the mosquito invasion was too important a story to ignore. His articles about the invasion generated nationwide attention, and, of course, the result was devastating to the Chrysler and Boeing recruiting programs. I cannot say with certainty how long this mosquito problem delayed MAF's staffing programs, but I can say with certainty that the program was indeed delayed. The pesky little critters proved to be a serious impediment to starting the manufacture of Project Apollo's first-stage boosters.

VIRGIL GRISSOM AND *APOLLO 1*

Astronaut Virgil "Gus" Grissom, a veteran of NASA's Project Mercury, Project Gemini, and soon to include Project Apollo, was assigned to MAF as part of NASA's Manned Flight Awareness program in early 1967. The flight awareness program assigned astronauts to the various NASA installations around the nation, which served a dual purpose. One purpose was to acquaint NASA and contractor employees with astronauts on a personal level, the result intending to instill a personal awareness that these rocket systems will be carrying humans to the moon. The other purpose was to provide the entire team of astronauts with updated reports regarding the status of rocket and spacecraft production at the various NASA operations throughout the country.

A couple of weeks before the disastrous rehearsal test of *Apollo 1* at Cape Canaveral's Launch Complex 34, I had to take Gus back from his visit to Michoud to his plane at Naval Station Belle Chasse across the Mississippi River from New Orleans to continue his journey to Cape Canaveral for the ground test of the *Apollo 1* spacecraft. As we all may recall, it was during that test when a cabin fire developed in the Command Module, a 100 percent oxygen environment—killing all three crew members: Command Pilot Virgil I. "Gus" Grissom, Senior Pilot Edward H. White II, and Pilot Roger B. Chaffee. The *Apollo* Command Module was destroyed. The decision was then made to change the Command Module's internal environment to a nitrogen-oxygen mix similar to our earthlike environment to ensure a safer spacecraft environment.

THE PRESIDENTIAL VISIT

On December 13, 1967, President Lyndon B. Johnson visited Michoud, and approximately twenty-six thousand visitors arrived to see the president. As always, the president's entourage included several secret service (SS) agents for his protection.

The plan was to have the president tour the facility traveling by golf cart, which was the usual mode of transportation throughout the enclosed, under-one-roof forty-three-acre building. The building's concrete flooring was sealed with an acrylic coating to ease keeping the floor clean. This coating, however, made the floor somewhat slick and caution had to be exercised while driving over it. I mention this because, as I had considerable experience driving the carts over the slick floor, it was determined that I should be the one who drives the president's cart. The president's cart was the lead cart with several additional carts attached to form a train arrangement. I also should mention that this was a time of considerable presidential assassination apprehension because the New Orleans district attorney Jim Garrison–Clay Shaw presidential assassination trial was in full swing. The idea that a president can be assassinated was a real issue, and it was accentuated by the Garrison trial. The president was well protected, however, with SS agents surrounding him in every direction—except the front, that is. His front was protected by "yours truly," a position I didn't particularly relish. Oh well, I figured it was all in a day's work.

As we were to begin the tour, the president was in a standing position directly behind me so that he could see better and be seen by the visiting public. I realized the danger of his standing while driving over the slick floor and felt I should ask him to sit down. Think about that—how does one tell, or ask, the president of the United States to sit down? I pondered about how I could politely do this, and in my haste to protect him and to begin the tour, I said, "Mr. President, would you please sit down? It is too dangerous for you to stand up during the tour." Immediately, of course, I not only realized that this was not the way to speak to the president but also realized that there is *no way* a president can be asked to sit down. Fortunately, he did not appear to be offended, and his reply to me was, "I'm alright, son. Let's get started." (BUT he continued to stand.)

The first half of the tour—through the Chrysler Space Division portion of the Michoud complex—went well, even with the president still standing directly behind me. As we approached the wall separating the Chrysler–Boeing portions of the site, a security cart preceding me along the route advised us to stop and await Louisiana governor John McKeithen who was running late but wanted to join the tour at that point.

While waiting for Gov. McKeithen to join us, and unbeknownst to me, one of the local television network affiliates laid a long, one-inch-thick TV cable directly across the front of my cart. After the governor arrived and we began to continue the tour, I began to move forward when my cart's front wheels grabbed the TV cable and caused the entire train to jerk forward. Because we were several carts linked together, all the carts bumped into each other, resulting in a loud gun-shot-sounding noise that threw everyone into a panic. The president, of course, fell forward and was within a couple inches to me. Immediately, the entire SS crew spilled on top of us in an attempt to protect the president. I don't think my welfare was given much thought at the time. Anyway, here we were, the president and me under a pile of people, and I wondered if the president had indeed been shot. I kept wondering if blood would start flowing from his mouth or nose, but fortunately that did not happen. Here I was within an inch or two from the president and for what it is worth, had the total attention of the leader of the free world. I immediately thought about how many worldwide leaders would wish they had this opportunity to say something profound to the president of the United States. This fleeting moment of fantasy passed, and all I could do was ask the president if he was okay. He said he thought he was okay, but to this day, I believe he also wondered if he had been shot.

When the gun-shot-sounding noise echoed throughout the enclosed forty-three-acre building, twenty-six thousand people simultaneously exclaimed, "Ooohhh," as they also thought the president had been attacked. Thinking this situation was my fault, I cannot describe to you the feeling one gets when something like this happens and twenty-six thousand people loudly exclaim their horror.

Somehow we all collected ourselves and completed the tour with the president giving a speech about the nation's effort to place men on the moon before the decade ends.

The evening news, particularly the *Huntley–Brinkley Report*, had much to say about the bumpy ride the president had in New Orleans. It was such an embarrassment to me that I don't think I spoke to anyone for two or three weeks.

SPANISH SECRETARY OF THE NAVY

My very first public speech/presentation occurred just a few weeks after joining NASA. The topic of my presentation was Project Apollo: How we planned to land on the moon and return our astronauts safely back to Earth.

My first audience was the Spanish secretary of navy who was visiting NASA facilities to better understand our mission. The secretary spoke limited English, so this complicated my presentation somewhat, having to have everything I said translated. Somehow I got through this event, and to my surprise, I was told the secretary was extremely pleased with the presentation and learned a great deal from it.

This experience, of course, gave me confidence in my speech-making ability and served me well for the many, many speeches required of me during my tenure with NASA. I even got to the point of enjoying giving speeches. Who would have thought that?

ON TO HUNTSVILLE

After several years at Michoud, I was transferred to the George C. Marshall Space Flight Center (MSFC) in Huntsville, AL, to serve in the MSFC's public affairs office. The Marshall Center was the parent agency of the MAF. This transfer was the first of two assignments to MSFC, having been assigned to the NASA–Mississippi Test Facility (MTF) in the interim to replace the MTF public affairs officer who was being transferred to NASA–Headquarters in Washington, DC.

It was the second assignment to MSFC that resulted in some of the more interesting experiences I encountered working for NASA. It was then and there that I was asked to serve as a special assistant to Dr. Wernher von Braun on various occasions. One can only imagine how thrilled I was to work directly with the man who, I believed, would be most responsible for the success of the US manned lunar landing program. Here was a guy so impressive it was hard to fully fathom. At age twenty-one, he was head of Peenemunde, the German research and development center (under Hitler) during World War II. It was here and then that von Braun's team developed the V-1 and V-2 rocket systems, the missiles used to attack Great Britain (it should be noted that von Braun resisted the use of his rocket systems for warfare, even to the point of being jailed because of his resistance. His sole objective was the peaceful exploration of space). It should also be noted that these rocket systems paved the way for usage of large liquid-propellant systems. Near the end of the war, von Braun and his team "surrendered" to the advancing US military under Operation Paperclip. This surrender was initiated by von Braun's brother Magnus.

To Jim Hennie
Many Thanks —
The Apollo 12 Crew
5-16-70

Apollo 11
Lunar Module on the surface of the moon

To
with warmest personal regards
Wernher von Braun

FOUR-FEET-TALL AMAZONIAN PYGMY TRIBAL CHIEF

One of the most unusual experiences while assisting von Braun was the time when the US State Department hosted a NASA-wide tour for an Amazonian tribal chief. The highlight of the tour was to be his introduction to Wernher von Braun.

Four feet tall, dark skinned, Prince Valiant-type haircut, and dressed in an all-white suit, the tribal chief eagerly awaited meeting the man who designed and developed the lunar landing launch vehicles. Translation, of course, was required, and von Braun was his ever-cordial self. I'll never forget that moment when I looked around the room and there I was with an Amazonian pygmy tribal chief, Dr. Wernher von

Braun, a State Department representative, and the late astronaut Pete Conrad who was a guest of von Braun at the time. I knew at that time that I probably would relate this apparently contradictory story for the rest of my life. And so I have.

An interesting sidebar to this event was the chief's inability to understand all the fanfare of flying to the moon. Keep in mind he had just flown from Brazil to New York, and, of course, you can't see New York from Brazil. But here, the moon is right there in clear view, so what is all the complexity about?

NIXON'S STATE DINNER TO HONOR THE *APOLLO 11* CREW

Another special assignment was to assist von Braun, his deputy Dr. Eberhard Rees—the gentleman who I always thought was the technical genius behind von Braun—and their wives while attending President Nixon's state dinner in Century City, CA.

Dr. von Braun was, of course, very much aware of his charisma, not at all in a vain sort of way but in a factual sense. He knew how his presence at a public function would draw attention, and he did not want to dilute in any way the attention that should be given to the *Apollo 11* crew. My primary job, therefore, was to insulate him from the media and promote full media coverage to the *Apollo 11* crew. I was given one exception though, and that was to allow Walter Cronkite to have access to him as they were personal friends. That situation did not arise, so von Braun was able to remain low-keyed.

In some ways though, he almost drove me crazy trying to make so many arrangements for him, e.g., make arrangements for him to pilot a seaplane. He was always going a hundred miles a minute, so yours truly had to somehow keep up with him.

The evening of the dinner, Mr. Rockwell of North American Rockwell Corporation hosted a reception for all dignitaries attending the event. Mr. Rockwell's reception was held at his all-glass-enclosed penthouse atop one of the Century City Towers, about a mile away from our hotel. This was a time of Vietnam War resistance, and our limousine traveled at a snail's pace through thousands of demonstrating hippies, carrying signs stating, "Make Peace, Not War." I recall Dr. von Braun turning to his deputy and saying, "Eberhard, we don't make war. We make love, don't we?"

After arrival at Mr. Rockwell's penthouse, I found myself completely overwhelmed. Even though I was only in my early thirties at that time, I truly believed I was somewhat sophisticated and capable of handling most situations. As it turned out, I learned I was not that sophisticated. Attending the reception were every who's who from Congress and industry—elephant tusks adorning the penthouse interior walls, mink throw rugs scattered about, and presidential helicopters flying in and out of the penthouse landing pad.

I suppose my overwhelmed state was obvious because in a matter of minutes, I got an elbow in my side. It was von Braun saying, "What's the matter, Jeem [his pronunciation of my name Jim]? Mingle, mingle!" I did my best to mingle, but unfortunately I never felt completely at ease.

My job seemed to never end at that event, and, of course, my not being a dignitary required my finding dinner wherever I could.

A KICK IN THE PANTS

Two of the more exciting occurrences happened while aboard the *Apollo 15* primary recovery ship, the aircraft carrier USS *Okinawa*, in August 1971.

The *Okinawa* was being escorted by a destroyer ship, and we were offered the opportunity to experience being transferred from a carrier to a destroyer while at sea. The method used was called a highline transfer using a Billy Pugh stretcher, which is normally used when it becomes necessary to transport an injured party to another ship for medical treatment. With both ships swaying and rolling in the wind and sea, it was an awesome experience for someone who was perfectly healthy and not in need of transporting. I couldn't help but wonder though what would happen to me if the cable used to transport me snapped. I think I know as I was being transported at a height of what seemed like a couple of hundred feet in the air. Still a great adventure.

Following the successful recovery of the *Apollo 15* astronauts, Commander David R. Scott, Lunar Module pilot James Irwin, and Command Module pilot Alfred Worden, I was asked to escort the large collection of moon rocks along with the collection of astronaut biological waste—a delicate description—back to the Johnson Space Center (formerly the Manned Spacecraft Center) in Houston, TX. I, of course, agreed but had some questions about how was I going to travel from an aircraft carrier to Houston, TX.

That experience was equally as exciting as it required me to be catapulted from the carrier deck and flown to American Samoa. For those who have never experienced being catapulted off an aircraft carrier, *it is truly a kick in the pants!*

Upon arrival at Pago Pago International Airport, located seven miles southwest of the central business district of Pago Pago, on the island of Tutuila in American Samoa, the first thing I did was head to the airport shower facilities and had what I felt was the most refreshing—and needed—shower of my life. The reason for this was after being at sea for a couple of weeks, freshwater for showering was becoming more and more limited. Seawater was added to the freshwater for showering, and this resulted in severe itching for the remainder of time aboard the ship. What a relief it was to be totally free of salt on my body.

Now that I was refreshed and ready to face the world again, we loaded the moon rocks and waste material on an Air Force C-141, a somewhat smaller version of the huge Lockheed C-5 Galaxy. There were only three of us aboard this huge aircraft, and we flew nonstop from Samoa to Houston, TX. The aircraft was so large we could have played basketball inside the fuselage.

After several hours of flight time, we arrived in Houston to a rousing crowd of several hundred people welcoming us back home. That experience, for a fleeting moment, made me feel like one of the astronauts returning home.

ONLY MY LAUNDRYMAN KNOWS

NASA announced the formation of the Mississippi Test Facility (now known as Stennis Space Center) on October 25, 1961, for testing engines for the Apollo program. A high-terrace area bordering the East Pearl River in Hancock County, MS, was selected for its location. NASA entrusted the US Army Corps of Engineers with the difficult task to procure each land parcel either by directly purchasing the land or through acquisition of a perpetual easement.

The initial requirements for NASA's proposed rocket testing facility required the site to be located between the rockets' manufacturing facility at Michoud Assembly Facility in eastern New Orleans, Louisiana, and the launch facility at the Kennedy Space Center in Florida. Also, the site required barge access as the rocket stages to be tested for *Apollo* were too large for overland transport.

The 13,500-acre test area is surrounded by a 125,000-acre acoustical buffer zone. The facility's large concrete and metal rocket propulsion test stands were used to test-fire the first and second stages of the *Saturn V* rockets.

Due to the severity of the low-frequency decibel level generated by the *Saturn* first- and second-stage engines, tests had to be conducted to determine under what environmental conditions would be best to conduct the tests. These loud sounds have been known to travel upward, then over long distances, and then return to Earth, causing considerable damage and consternation to an unsuspecting public.

A large acoustical horn was built to simulate the noise level generated during an actual test-firing. The horn would be turned on during certain environmental conditions to test its impact.

One day while trying to acclimate myself to my new assignment, I wandered over to the acoustics lab to familiarize myself with its operation. Unbeknownst to me, the horn was located atop the lab building, and as fate would have it, it was turned on as I was walking by. The physical impact of that moment was nothing like I had ever experienced before. It was as though I was standing in a vibration booth with sound waves bouncing off my body in an uncontrolled manned. I've often said, "Only my laundryman really knows the impact that horn had on me!"

MY FRIEND AND COWORKER JUST DOESN'T LOOK RIGHT!

Upon completion of our *Apollo 12* assignments aboard the USS *Hornet*, my friend and coworker Robert (not his real name) took the first flight home we could find. As news about our involvement with the mission became known aboard the aircraft, we were treated like celebrities by the plane crew and passengers alike. We weren't sure how to handle this celebrity treatment but enjoyed it nevertheless. The stewardesses—as they were then called—went way out of their way to accommodate us, probably too far out of their way. We could hardly finish a libation when another one was placed on our tray. I have no recollection of how many drinks we both consumed during that several-hour flight, but it was substantial.

Upon our arrival in Dallas, TX, where Robert and I would separate and continue on to our respective destinations, we thanked everyone for their generous treatment of us and disembarked the plane. After cordial good-byes, Robert proceeded down the concourse to his next flight location. As he was walking away, I could not help but notice he did not look quite right. Still impeccably dressed—sport coat, dress shirt, tie, and dress slacks—Robert continued on his way, but I still felt something was not right. After closer inspection as he continued walking away, *I discovered he was not wearing his socks and shoes.* Fortunately, we were able to get him back to the airplane, and the stewardesses went back aboard and found his socks and shoes. We somehow convinced him that his appearance would be so much improved if he completed his sartorial splendor. He reluctantly agreed, and once again, he was on his way to the final leg of his journey.

WORKING THE APOLLO MISSIONS

I was assigned to the news desk at the NASA–Johnson Space Center, Houston, TX, for all but the last few Apollo missions. For *Apollo 12* and *Apollo 15*, I was assigned to the Primary Recovery Ships (PRS) as the NASA coordinator of radio and television. My job there was to serve as a NASA coordinator for the navy and television/radio network pools. Interesting job, keeping both the navy and network pools happy by satisfying their news coverage requirements.

Assignments to the recovery ships had it perks, mainly the opportunity to visit Hawaii before shipping out. My wife joined me on the trip to Hawaii for *Apollo 15*, and I recall her last comment to me when she departed Hawaii. She was aware of my interest in snorkeling in the nearby beautiful clear waters, and because we both heard the stories of visitors getting robbed while snorkeling away from their vehicles where their wallets and clothing were stored, her parting words to me were, "Be careful. Don't get robbed!"

I don't suppose I have to tell you the rest of the story. Needless to say, she was able to wire some additional money to me before I had to board the ship.

One of the highlights on the ship was getting up early one morning—still dark outside—and seeing the reignition of the *Saturn V*'s third stage (one-hundred-plus miles above in Earth's orbit), sending the spacecraft on its way to the moon. This reignition occurs after the third-stage *Apollo* spacecraft is placed into Earth's orbit—called parking orbit—in order to check out the system after the rigors of lifting off from Earth. After the system checkout, the spacecraft was then sent on a path to rendezvous with the moon a couple of days later.

I ENDED UP A BLOODY MESS!

Apollo 12 was my first deployment to a Primary Recovery Ship. The USS *Hornet* served as the PRS for the first lunar landing mission—*Apollo 11*—and was selected again as the PRS for *Apollo 12*.

The *Hornet* departed Pearl Harbor several days before the actual liftoff date from Cape Kennedy—November 14, 1969—in order to be positioned at various locations in the Pacific in case there was an early termination of the flight. This was typical protocol for all *Apollo* missions. We sailed to the predetermined locations; fortunately, our services were not needed as the mission was given the go-ahead to land on the moon. This early departure also provided time for the recovery teams to practice their recovery techniques. On board the

Hornet was a full-scale mockup of the command module (CM), the top part of the spacecraft that housed the three astronauts at liftoff and reentry into the earth's atmosphere at the end of the mission. The CM mockup would be dropped off into the Pacific; the ship would continue to sail away from the mockup and then turn around to begin the recovery activity. If one was not aware of what was going on, he or she would think they were witnessing the actual recovery operations. By the time splashdown arrived, the deployed recovery team was as proficient as could be hoped for.

Another activity that I was totally unaware of was being planned as our ship approached the equator. I had never heard of the Kingdom of Neptune or the Shellback Initiation, which was a requirement for all individuals crossing the equator for the first time. I was told that individuals crossing the equator for the first time in a commercial aircraft would be forced to drink a cup of champagne. Hearing this, I reasoned that the initiation aboard the *Hornet* would not be that bad.

Well, was I wrong! The Shellback Initiation was a longtime naval tradition that initiated pollywogs into the Kingdom of Neptune, the legendary god of the seas. I later learned that skippers of ships have always taken time, either in war or peace, to uphold this long-practiced ceremony. We were told that the *Hornet*'s captain had given orders that we NASA civilians were not to be treated too harshly; however, I don't think the word was passed on.

To prepare for the ceremony, equator-crossing veterans would pile up stores of garbage—days-old food garbage—smelly clothing and anything else that is vile but not life threatening. That last part was good to know.

I tried to learn as much as I could about the initiation so that I could prepare myself as much as possible. Having learned that we would have to crawl across the pumice flight deck, I padded my knees with bubble-wrap, which would be hidden by the jeans I would be wearing, hoping it would offer some protection.

On the appointed day and time, the uninitiated, yours truly included, were led out onto the flight deck. As none of us was wearing any kind of badge or other designation to distinguish us as civilians, we were regrettably integrated into the ship's crew that was to be initiated. Any thoughts of lenient treatment went out the window.

As we all were treated in a humiliating manner, we were made to run—actually crawl—a gauntlet, being belted with greased socks, parachute straps, and whatever else the initiators could find. My bubble-wrapped protective knee coverings held up for, I'm guessing, maybe three or four feet at the most. In short order, my knees were a bloody mess. At the end of our crawl-space, we were placed individually into a coffin-sized wooden box, which contained the previous week's rotting food items, and made to roll around the slop to ensure our getting completely contaminated. To this day, I can recall/smell that horrible stench.

Following the initiation, we could not wait to get into a shower. As I was showering, my comrades noticed and called to my attention how I was bleeding not only from my knees but from my back and sides. I decided to not be a sissy, so I accepted my bruised and bloody situation. Not happy about it, but what choice did I really have.

On a positive note, it's safe to say I will forever treasure the Kingdom of Neptune certificate that was given to us after the initiation. The certificate reads as follows:

TO ALL SAILORS WHEREVER YE MAY BE: *and to all Mermaids, Whales, Sea Serpents, Porpoises, Sharks, Dolphins, Eels, Skates, Suckers, Crabs, Lobsters, and all other Living Things of the Sea.*

GREETINGS: Know ye that on this day in November, 1969, there appeared within Our Royal

*Domain the U.S.S. Hornet bound for the equator and for **Espiritu Santo in the Hebrides Islands**.*

***AND BE IT KNOWN** By all ye Sailors,*
Marines, Land Lubbers and others who
may be honored by his presence that

James A. Henrie

Having been found worthy to be
numbered as one of our Trusty Shellbacks
he has been duly initiated into the
SOLEMN MYSTERIES OF THE
ANCIENT ORDER OF THE DEEP

As you may guess, there were many more exciting moments during my NASA years, but the above are those events that hold special significance for me. I will forever be grateful for the wonderful anecdotal snippets of my life within NASA.

Made in the USA
Monee, IL
04 November 2026

67375609R00036